The Library of Subatomic P

The Neutrino

Fred Bortz

The Rosen Publishing Group, Inc., New York

To Susan, whose quiet influence is cosmic to many

Published in 2004 by The Rosen Publishing Group, Inc.
29 East 21st Street, New York, NY 10010

First Edition

Library of Congress Cataloging-in-Publication Data

Bortz, Alfred B.
The neutrino / Fred Bortz.
    p. cm. — (The library of subatomic particles)
Summary: Tells how the neutrino, an almost massless subatomic particle without a charge, was discovered and helped to explain the process of radioactive decay.
Includes bibliographical references and index.
ISBN: 978-1-4358-9160-9
1. Neutrinos—Juvenile literature. 2. Particles (Nuclear physics)—Juvenile literature. [1. Neutrinos. 2. Particles (Nuclear physics)] I. Title. II. Series.
QC793.5.N42B68 2004
539.7'215—dc22

                                    2003015112

Manufactured in the United States of America

On the cover: An artist's illustration of quarks making up protons and neutrons, and the exchange particles, the pi mesons and gluons, that hold them together.

Contents

Introduction

You have probably learned that matter is made of atoms and that atoms are made of protons, neutrons, and electrons. But not all subatomic particles are contained in matter as you are used to thinking about it. This book is about one of those other particles, the neutrino, whose name is Italian for "little neutral one." In the chapters ahead, you will discover that although a neutrino doesn't carry any electric charge, has very little mass, and is very hard to detect, understanding its story is very important for making sense of the entire subatomic world.

Neutrinos are almost surely the most numerous of all subatomic particles. The mass of a neutrino is so small that no one has yet succeeded in measuring it. But if you could collect all of them in the universe, their mass would probably be as great as all the protons, neutrons, and electrons in all the galaxies and the vast dust clouds between them. Neutrinos stream outward from the Sun and other stars, and billions of them zip through your body every second. Yet most of those neutrinos pass through Earth without affecting even a single atom. It takes careful scientific work to detect them. In fact, it took about twenty-five years from the time physicists (scientists who study matter and energy) first realized that neutrinos were necessary in nature's scheme of things until a team of researchers identified them in a nuclear reactor.

To understand neutrinos, you first need to understand atoms and their parts, especially the nucleus and the phenomenon of radioactivity. The nucleus was discovered less than a century ago, and physicists are still exploring the forces within it. This book follows their paths of discovery to the modern limits of particle physics, where many of the most important news stories come from those nearly massless uncharged sprites known as neutrinos.

Chapter One

Atoms and the Forces of Nature

Physicists often notice patterns in natural phenomena and look for the fundamental principles underlying their observations. Sometimes patterns seem complex, until someone finds a new viewpoint from which things appear to be simpler. That viewpoint often leads to important discoveries about the laws that govern nature. To help you understand the way physicists think, let's start not with the unfamiliar forces deep within atoms but with a force you experience all the time—gravity.

To people of ancient civilizations, Earth was unmoving and was the center of everything. The Sun, Moon, and stars seemed to follow regular paths around it, day after day, year after year. Everything in the heavens followed easily predicted paths—except for the planets. People recorded their motions across the sky and

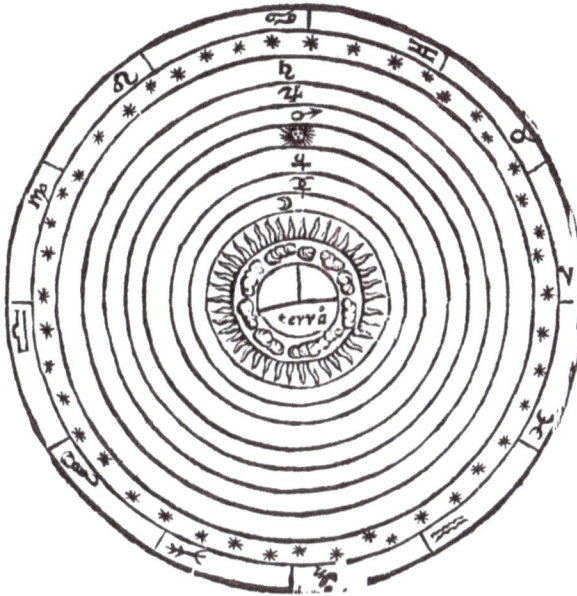

SYSTEMA PTOLEMAICVM.

terra

B 2   Syfte.

Living in the Center of the Universe. This diagram shows the interpretation of the skies by the renowned second-century mathematician, astronomer, and geographer Claudius Ptolemy, who placed an unmoving Earth at the center of the universe, with everything else going around it daily in circular paths. Progressing outward from Earth are the Moon, Mercury, Venus, the Sun, Mars, Jupiter, Saturn, and the stars, including the constellations of the zodiac.

discovered that they would sometimes speed up, slow down, or even reverse their direction of movement. The planets weren't following a simple path around Earth like the Sun, Moon,

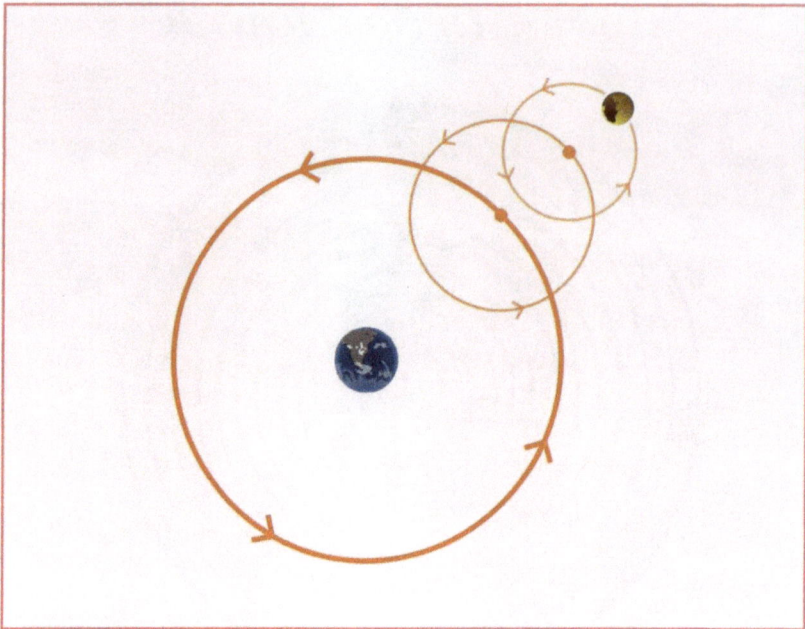

Circles Upon Circles. Because the planets were known to move in different directions at different rates and different times, Ptolemy needed to add some complications to his diagram. He devised a system of epicycles, or circles whose centers moved on other circles, that seemed to explain their paths. But as astronomical measurements improved, other astronomers needed to add epicycles to epicycles.

and stars, so the sky watchers had to do a little tinkering with the natural laws to make planetary motion fit. They began with a simple modification: suppose each planet's path consisted of a small circle called an epicycle whose center went around Earth in a larger circle. That matched observations fairly well, but as they gathered more detailed measurements, the epicycles failed to match the planets' movements. Soon they began adding epicycles to epicycles and things were beginning to look too complicated to be sensible.

A Simpler View. By the sixteenth and seventeenth centuries, astronomers like Nicolaus Copernicus realized that the complicated system of epicycles was unnecessary if Earth was a planet like the others, in orbit around the Sun. Johannes Kepler, shown here, applied his mathematical talents to the best measurements and concluded that the orbits were slightly oval rather than perfectly circular.

Some people found a different way of interpreting planetary motion, with the Sun at the center of the solar system and Earth and all the other planets in orbit around it. That got rid of many epicycles, but not all, and for more than a thousand years, most people were reluctant to accept that their world was not the center of everything. Finally, Polish astronomer Nicolaus Copernicus (1473–1543) wrote a book that persuaded

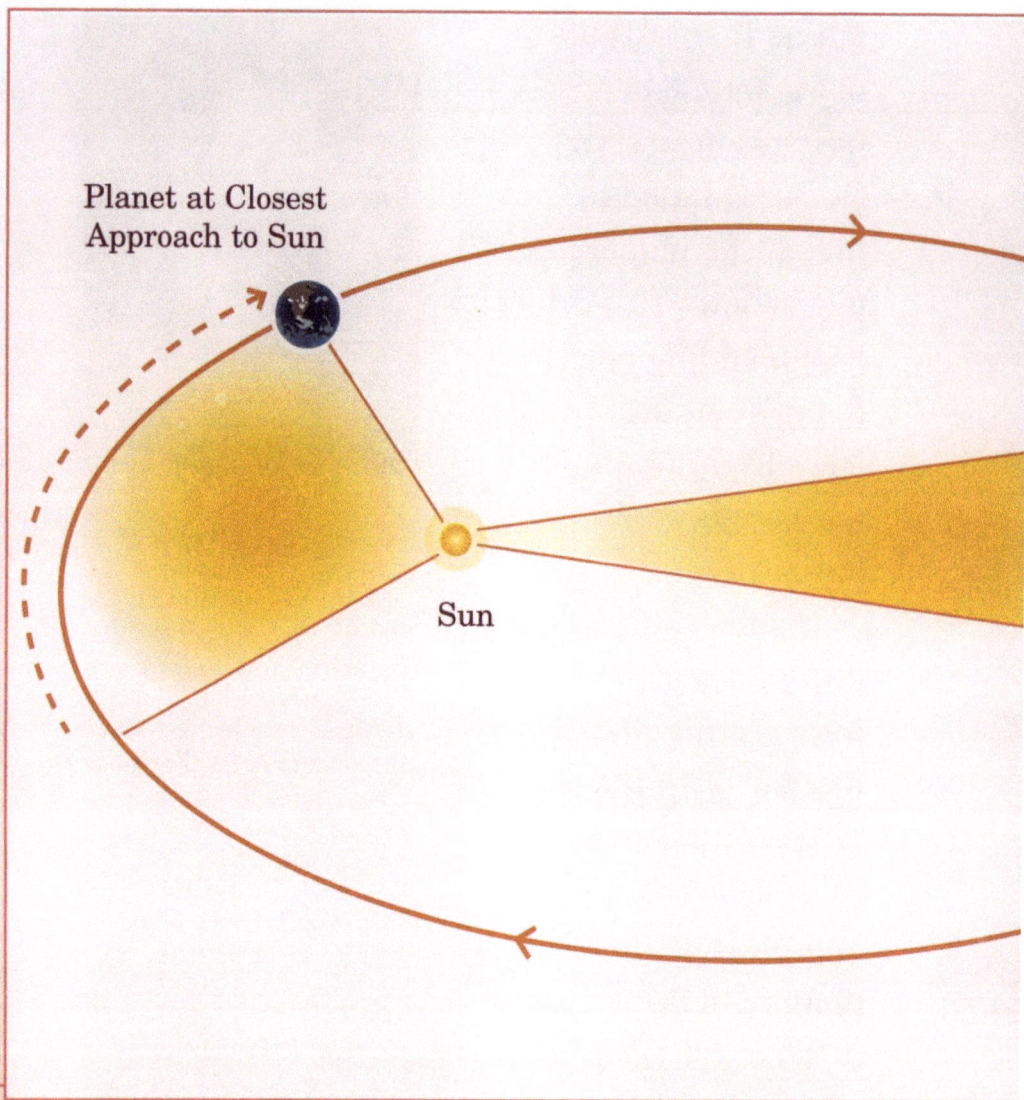

Planet at Closest
Approach to Sun

Sun

Kepler's Laws. This diagram shows two of Kepler's three laws of planetary motion. His first law states that an orbiting body like a planet follows a path that is not a perfect circle but rather an oval shape called an ellipse. (The ellipse in this diagram is exaggerated—most orbits are almost circular.) Another of Kepler's laws states that a planet moves faster when it is close to the Sun and slower when it is farther away, but the line from planet to the Sun sweeps out the same area (shaded) in the same length of time.

Planet at
Farthest
Point
from Sun

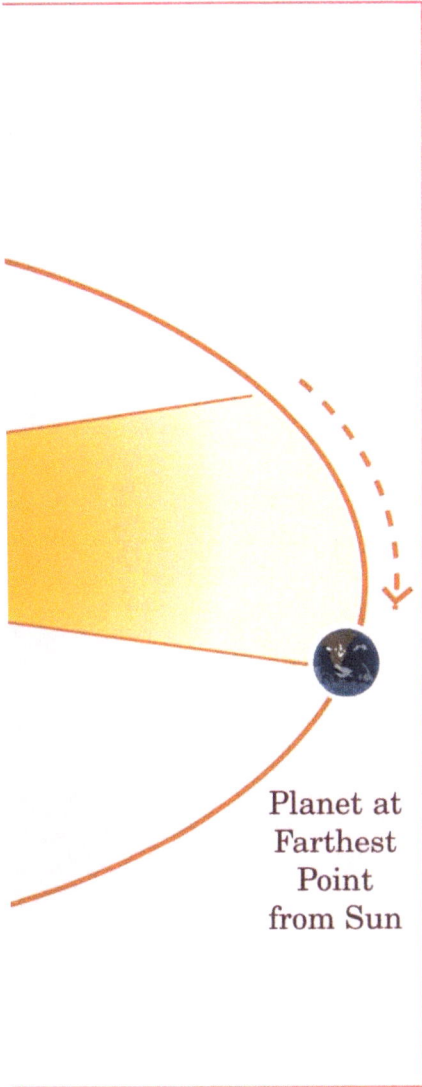

many people that Earth was a planet like the others. Copernicus's idea finally took hold when German astronomer Johannes Kepler (1571–1630) applied his mathematical talents to the best astronomical measurements of his time. He found that the planets moved not in circles, but in curved paths called ellipses. He also discovered two other mathematical formulas that described the details of the motions of planets in their orbits. Thanks to Kepler, nature was once again governed by simple laws. Still, those laws were only a description of planetary motion, not an explanation of the underlying cause.

That explanation came from Sir Isaac Newton (1642–1727), who proposed that the laws of nature are the same everywhere. The same gravitational force that attracts apples downward from trees also draws the Moon toward Earth and the planets toward the Sun. Newton also developed three laws of motion that scientists and engineers still rely on today. The first two laws describe how an object's motion changes when a force acts on it. The third law states that any interaction between two bodies consists of equal forces acting in opposite directions on them.

Newton expressed those laws as mathematical formulas and used them to analyze the motion of a smaller body orbiting a much larger one, such as the planets going around the Sun. Out popped elliptical orbits and Kepler's other two formulas. Thanks to Newton, gravity was recognized as a universal force. His third law of motion also established an important principle that would lead to the discovery of the neutrino centuries later. Because the forces on two interacting bodies are equal and opposite, so are their changes in momentum (a quantity that measures the direction and intensity of motion). Thus, the total momentum before, during, and after the interaction remains unchanged. That principle, known as conservation of momentum, has been a pillar of physics for more than three centuries.

Chemistry, Atoms, and Electromagnetism

In the nineteenth century, scientists studied other forces that might also be universal, particularly electricity and magnetism, which were discovered to be different aspects of the same phenomenon called electromagnetism. Meanwhile, English meteorologist John Dalton (1766–1844) had begun studying the gases of the air, hoping his research would help him to understand more about weather. He ended up transforming the science of chemistry instead, by reviving an ancient idea that all substances were made of indivisible bits called atoms. In his 1810 book called *A New System of Chemical Philosophy*, Dalton wrote about elements, substances made of only one kind of atom, and compounds, substances made of specific combinations of atoms called molecules.

Dalton observed chemical reactions to determine how different elements combine with each other and to calculate their atomic masses. He set the atomic mass of hydrogen, the lightest element, at one unit and based the mass of other atoms on that. Dalton also stated an important principle about the way matter could change in chemical reactions. The atoms might reshuffle in their combinations, but they never change from one element to another. As a result, the mass before the reaction was the same as the mass afterward, a principle known as conservation of mass.

As chemists identified more compounds and the elements that composed them, they gradually developed a list of the atomic masses and other properties of each element. By the end of the 1860s, scientists knew of sixty-three elements and could see distinct similarities and differences among their properties. They could group elements according to patterns in their atomic masses, their melting or boiling points, their densities (the mass of each cubic centimeter), the way they combined with other elements, and the properties of the compounds they formed. Still, no one had come up with a successful arrangement of the elements that would enable people to see all these similarities and differences on a chart. Such a chart might also reveal an underlying principle of matter.

Early in 1869, Dmitry Ivanovich Mendeleyev (1834–1907), a chemistry professor at St. Petersburg University in Russia, was taking a long train trip and occupying himself by arranging and rearranging a deck of home-made cards, each listing an element and its properties, as if playing solitaire. He finally discovered a pattern that made sense. Starting from the smallest atomic mass and increasing as he went along, he laid out a column of cards. He continued until he came to an element that had the same chemical property, or valence, as one of his earlier cards. Then he would start a new column and repeat the process, producing rows of elements with the same valence. An element's valence describes how it tends to form compounds with other elements. For example, the alkali metals—lithium, sodium, potassium, rubidium, and cesium—all form compounds of one atom each with nonmetallic elements known as halogens—fluorine, chlorine, bromine, and iodine. The alkali metals have a valence of +1 and the halogens have a valence of -1, so the compound has a net zero valence. Valence also relates to electrical properties of the elements, as chemists were discovering by passing electricity through compounds dissolved in water.

Because the patterns of similar valence repeated, Mendeleyev called his arrangement the periodic table of the elements. It is still in use today, though with the elements in rows and the valences in columns. The arrangement had gaps, but that was the most important part of Mendeleyev's success. He recognized that the gaps represented undiscovered elements. He boldly predicted not only which elements would be found but also what their atomic masses and densities would be, and he was right! The periodic table provided powerful evidence that atoms, though not yet observed directly, were real. But why was it periodic and how was that related to valence? It took more than sixty years of research to find the explanation, and part of it was because Dalton was not quite correct. He was right that atoms are the smallest bits of an element, but he was wrong to consider them indivisible.

Inside the Atom

As the nineteenth century was ending, two major discoveries provided hints that atoms were made of smaller pieces. The first was the phenomenon of radioactivity, and the second came from the study of cathode rays. In February 1896, French physicist Antoine-Henri Becquerel (1852–1908) was studying substances that glow in the dark after being exposed to sunlight, such as compounds of the rare metal uranium. He soon discovered that they produced radiation even without being exposed to sunlight. In the next few years, chemists and physicists discovered a number of radioactive elements, and laboratories all

Discoverer of Radioactivity. In 1896, Antoine-Henri Becquerel, looking for materials that glowed in the dark, discovered that compounds of uranium produced rays that would darken photographic film, even without being exposed to light. The phenomenon soon became known as radioactivity.

around the world began to study the nature of radioactivity. By then, physicists had developed another important principle, conservation of energy, to go along with conservation of mass and momentum. Energy might change form, such as from the heat of steam to the mechanical motion of a piston, but it could not be created or destroyed. Where was this radioactive energy coming from?

In 1895, just as the news about X-rays sent physicists scurrying to learn more, a promising young student from New Zealand, Ernest Rutherford (1871–1937), arrived at the Cavendish Laboratory of Cambridge University in England. Lab director J. J. Thomson (1856–1940) promptly assigned

Rutherford the task of understanding the new rays and then extended the assignment to include radioactivity when Becquerel announced his surprising results.

Meanwhile, Thomson was busy with the final stages of a series of experiments to understand the cathode rays seen in glass tubes from which most of the air had been removed. The terminals of a battery or electric generator were connected to two electrodes in that tube. The cathode, or negative electrode, was a heated metal filament, and it emitted a beam, called a cathode ray, that caused a glow. Thomson's objective was to find out what that beam consisted of. In 1897, he announced his results. Cathode rays were tiny particles less than one-thousandth as massive as a hydrogen atom but carrying as much negative electricity as the hydrogen nucleus carries in positive charge. He called the particles corpuscles and concluded that they were particles from within the atoms of the metal. Today we know them as electrons, and we know that they are responsible for the chemical behavior of every substance.

Might radioactivity also come from inside atoms? Rutherford began to answer that question in his three years at the Cavendish Laboratory, where he discovered that radio-activity had two distinct forms, which he named alpha and beta radiation after the first two letters of the Greek alphabet. He continued to study them when he became a professor at McGill University in Montreal, Canada, in 1898. Each of those forms of radiation is important to our story. As you will read in the next chapter, alpha radiation would lead Rutherford to discover the atomic nucleus, and beta radiation would eventually lead to the prediction of the existence of the neutrino—but not before setting off alarm bells about physicists' treasured conservation laws.

Chapter Two

Neutrinos to the Rescue

When Rutherford studied radioactivity from uranium at the Cavendish Laboratory, he first discovered that it had two distinct components—one that was easily blocked by a few layers of aluminum foil and one that was much more penetrating. He called them alpha rays and beta rays. In 1898, he accepted a professorship at McGill University with one of the world's best physics laboratories. By 1900, he had discovered a third even more penetrating form of radiation, which he called gamma rays.

Rutherford had also found hints that radioactivity violated one of Dalton's fundamental principles by transforming one element into another. Radioactive thorium was producing an "emanation," a radioactive gas, which we now know as radon. He needed a chemist to analyze the changes and found the perfect

Discoverer of the Nucleus. Ernest Rutherford was among the first people to study radioactivity. He and his colleagues discovered three distinct types of radioactivity, which they named alpha, beta, and gamma, after the first letters of the Greek alphabet. He then went on to explore the structure of atoms and discovered the nucleus.

colleague in Frederick Soddy (1877–1956), who had just arrived at McGill. Together, Rutherford and Soddy tracked the elements from their original form to their new forms. They found that when an atom emits an alpha particle, its atomic mass decreases by four units and its atomic number (its position in the periodic table) decreases by two. When it emits a beta particle, its atomic number increases by one unit but doesn't change atomic mass. They recognized that this wasn't a chemical reaction but a "transmutation," the transformation of an atom of a "parent" element into the atom of a different "daughter" element. The daughter is frequently more radioactive than its parent, so there is a continuing sequence of transmutations—a chain of alpha or beta decays—from one atom to another to another.

Discovery of the Nucleus

Rutherford and Soddy were clearly on the track of something important. Becquerel had already identified beta rays as lightweight, negatively charged particles—electrons, in fact. Rutherford and Soddy likewise found that alpha rays were much heavier positively charged particles. Transmutation results suggested that an alpha ray was a helium atom without its electrons. Today we call that a helium nucleus, but the atomic nucleus was still to be discovered—by Rutherford, of course. That great discovery took place back in England, at the University of Manchester, where Rutherford accepted a professorship in 1907.

At Manchester, Rutherford used alpha particles as bullets, directing a beam of them at thin metal foils. If he could measure their scattering, or the pattern of their deflections, from the atoms in the foil, that might reveal the

Radioactive Source

Alpha Particles

Detector

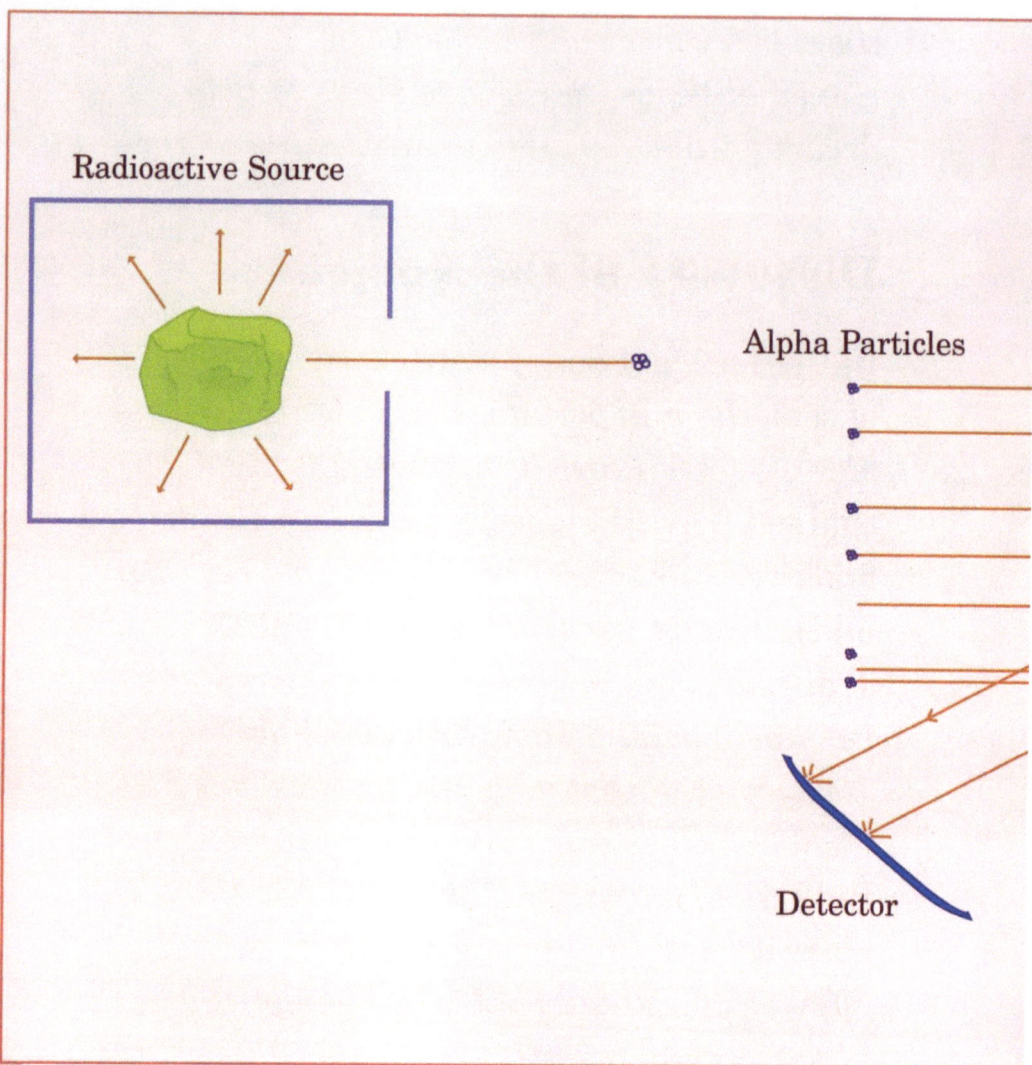

Scattering Experiments. With his students Hans Geiger and Ernest Marsden, Rutherford used alpha particles as bullets and detected the way they deflected, or scattered, from thin foils of metal. They discovered that most alpha particles went through without being deflected much at all, but a few, much to the scientists' surprise, went far off to the side or even bounced back toward the source. Rutherford interpreted the result to mean that most of the mass of an atom is concentrated in a very tiny central body called the nucleus.

Thin Strip of
Metal Foil

Detector

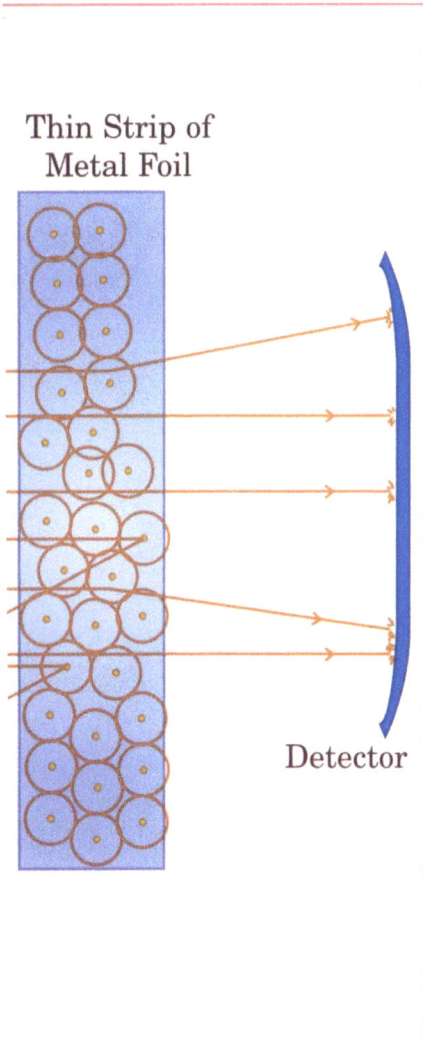

atoms' size, spacing, and perhaps even their shape or internal structure. His student Hans Geiger (1882–1945) devised an instrument that detected and counted alphas, and in 1909, he began the scattering experiments. Nearly all the alphas passed straight through the foil or deflected only slightly. That fit with J. J. Thomson's idea that atoms were like plum puddings, with the atomic number of electrons scattered like tiny plums in the pudding's bulk, which had nearly all the mass and enough positive electric charge spread out to balance the negative electrons.

But Geiger's experiments produced a puzzle. His

detectors were very accurate, but a few alpha particles were unaccounted for. Had they scattered beyond the detectors? If so, what could cause scattering through such large angles? Rutherford thought the task of looking for large-angle scattering would be good experience for Ernest Marsden (1889–1970), a young student just learning the techniques of research. Marsden found the missing alpha particles. Some went to the left or right of the original detectors, and a few even scattered backward!

By 1911, Rutherford had an explanation for the results. Atoms, he stated, were like miniature solar systems, mostly empty space and held together by electricity instead of gravity. Tiny electrons are the planets, carrying only a small fraction of the system's mass. The central body, called the nucleus (plural: nuclei), though much more massive, is very compact, occupying about one ten-thousandth of the diameter of the atom. The emptiness of the atom explains why most alpha particles pass through with little scattering. But on those rare occasions when a fast-moving alpha particle makes a nearly direct hit on a heavy nucleus, the alpha scatters sideways or even backward.

Rutherford then set out to understand what made one element's nucleus different from another. In 1919, he replaced the retiring J. J. Thomson as leader of the Cavendish Laboratory at Cambridge University. By then, physicists accepted Rutherford's idea that the nuclei contain a number of protons, particles that carry the same amount of positive electric charge as the electron's negative charge. They also agreed that the atomic number of an element was the number of electrons it had and that its nucleus contained a balancing amount of positive charge from protons. Thus, the nucleus of hydrogen was a single proton and alpha particles were helium nuclei with two units of positive electrical charge and an atomic mass of four. Where did the extra atomic mass units come from? For larger atoms, the discrepancy between atomic number and atomic mass was even greater. Lead, for instance has atomic number 82 and atomic mass 207. Rutherford proposed that the extra mass came from electrically neutral subatomic particles with about the same mass as protons. He called them neutrons. Alpha particles, for example, consisted of two protons, each with one unit of positive electric charge, and two uncharged neutrons. Not all physicists agreed until neutrons were detected in 1932.

Relativity, Quantum Mechanics, and Conservation Laws

Rutherford had correctly explained that alpha decay occurs when a stable helium nucleus bursts out of a larger unstable nucleus. His explanation of beta emission, however, was not quite correct. He said that it resulted from the splitting of a neutron in an unstable nucleus, producing a proton and an electron. That seemed sensible, until physicists studying the energy of beta particles ran into problems with conservation laws.

The early decades of the twentieth century were full of major surprises for physicists. Probably the most famous of these was the theory of relativity, devised in 1905 by Albert Einstein (1879–1955). One of its most surprising predictions—even to Einstein himself—was that mass and energy are two aspects of the same phenomenon, as represented by the formula $E = mc^2$. The amount of energy (E) in a unit of mass (m) can be determined by multiplying the mass by the speed of light (c) times itself (or squared). That formula explains the source of the energy carried off by alpha or beta rays. As scientists made more careful measurements of the mass of parent and daughter nuclei in

A Surprising Formula. When Albert Einstein, shown here, developed his theory of relativity, he was surprised that it led to an equation, $E = mc^2$, that showed that mass and energy are two forms of the same quantity. When scientists applied that result to the phenomenon of radioactivity, they understood where the energy was coming from.

radioactive decay, they discovered that the mass of the daughter element plus the emitted alpha or beta particle was less than the mass of the parent. When that missing mass was put into Einstein's formula, the result matched the emitted alpha particle's kinetic energy (energy of

motion). For all alpha decays from a particular parent nucleus, the alpha particle always carried off the same amount of energy.

For beta emission, the results were less satisfying. The beta particles from a parent nucleus had a range of kinetic energies, from very little to a maximum amount that fit Einstein's equation. Where was the missing energy of the slower beta particles? Not only that, but a new understanding of the subatomic world had been developing since Rutherford proposed his planetary picture of the atom, and it, too, caused problems with beta decay. Called quantum mechanics, it had revealed that a particle like an electron is described by a set of properties that can only have certain values. Collectively these properties are called its quantum state. One of these properties describes its magnetic alignment by a quantity called spin. Spin, like energy, momentum, and electric charge, is conserved in any interaction of particles. For protons, neutrons, and electrons, the spin values are the same (1/2 of a natural quantity known as Planck's constant), but they may have either a positive or negative sign.

Neutrino

Neutron About
to Decay

→ x

Proton

Electron

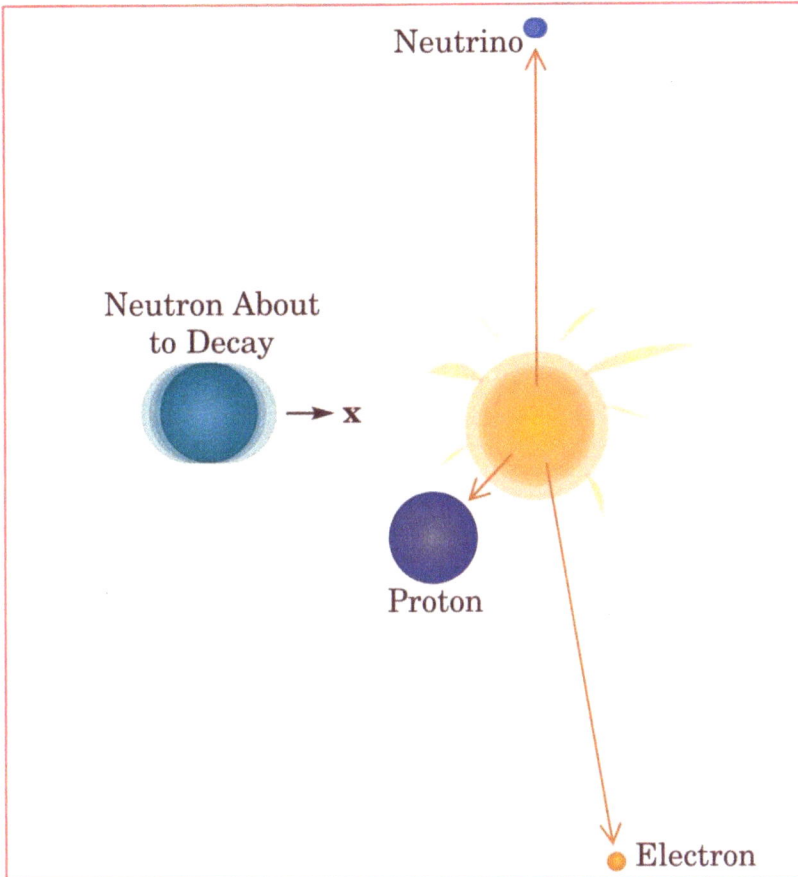

The First Hints of Neutrinos. Measurements of the energy released in beta decay at first puzzled physicists. They understood that a neutron could become a proton by emitting an electron (the beta particle), but the electron's energy could take on any value from zero up to the full amount of mass lost in the transformation. Believing in the well-established laws of conservation of energy, momentum, and electric charge and the emerging law of conservation of spin, in 1933 Enrico Fermi proposed that each beta particle was emitted with a tiny neutral partner called a neutrino.

Beta emission begins with a neutron, so the total spin before decay is either +1/2 or -1/2. If the only products of the emission are a proton and an electron, the total spin afterward is +1, -1, or 0. The total electric charge is zero before and after beta emission, but what can be done to conserve spin? In 1930, Austrian physicist Wolfgang Pauli (1900–1958) suggested that the beta particle must be emitted with a partner. That partner would have a spin of 1/2, but would be electrically neutral and have little or no mass. In 1933, physicist Enrico Fermi (1901–1954) developed a theory of beta decay that incorporated what he called the "little neutral one," or neutrino in his native Italian. Its lack of charge and low mass would make it very hard to detect, but all signs pointed to its existence. Finally, in 1956, using measurements made inside a nuclear reactor where there were lots of neutrinos flying around, physicists managed to detect a few. The neutrino had come to the rescue of the theory of beta decay—and it had some more heroics yet to perform in the subatomic world.

Chapter Three

The Particle Zoo and Neutrino Number Two

Quantum mechanics also led to a new scientific understanding of what a particle really is. It seems natural to think of particles as entities with definite places, shapes, and sizes, and it seems natural to view waves as different kinds of entities that are spread out and moving in space and time. But at the subatomic level, waves and particles are not distinct at all. Quantum mechanics provides a different picture of electrons in atoms than the one in Rutherford's model. Instead of being like planets moving in orbits, the electrons are like tiny vibrating violin strings with their ends tied together to make that orbital shape. Likewise, quantum mechanics describes light and other electromagnetic energy as flowing not in perfectly smooth waves but rather as grainy streams of particles called photons.

Intensity

Time

Intensity

Time

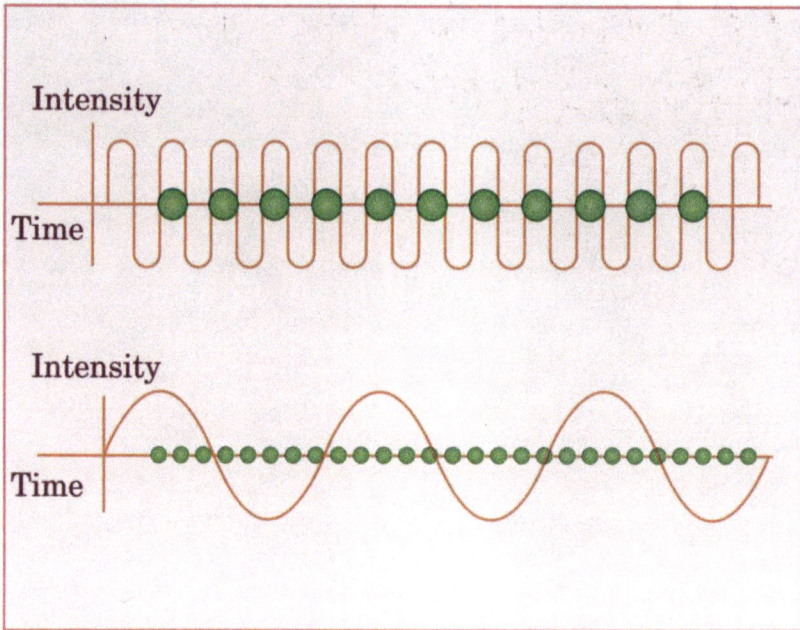

Grains of Light. Quantum mechanics led to a new understanding of electromagnetism. Light and other electromagnetic waves now had to be viewed under some circumstances as being a grainy stream of particles called photons, whose energy depended on the frequency of the wave. Likewise, when charged subatomic particles interacted electrically or magnetically, the force was understood to be the result of the exchange of photons.

Though blending waves and particles seemed as odd as discovering that mass and energy are the same thing, physicists quickly accepted both quantum mechanics and relativity because of their extraordinary success in describing a wide variety of atomic and subatomic phenomena. Light really is grainy, electrons really can act like waves, and the energy of radioactivity comes from transforming some mass. But that meant some very successful older theories of physics, like the equations of electromagnetism that produced smooth electromagnetic waves, needed to be modified at the subatomic level. Many physicists struggled to come up with a new description of electromagnetism, which they called quantum electrodynamics, or QED for short. Although QED was not fully developed until the 1940s, physicists made a good start on it in the 1930s when they described electromagnetic forces between electrically charged particles as the result of exchanging photons between them. If the particles had opposite electric charges, the photon exchange produced an attraction. If they had the same charge, the photon exchange would lead to repulsion (pushing apart). In QED, the photon is the carrier of the electromagnetic force.

Naturally, physicists wondered if a different exchange of particles might explain how the nucleus holds together despite having all those electrically repelling protons clustered together in such a small space. Could protons and neutrons be exchanging particles, too? In 1935, Japanese physicist Hideki Yukawa (1907–1981) developed a theory suggesting that nuclei hold themselves together by a force we now call the strong nuclear force, or simply the strong force. It results from exchanging particles known as pions, each with a mass about 250 times that of an electron. Another force within the nucleus, the weak force, explains beta decay, because it keeps a neutron from transforming into a proton, electron, and neutrino most—but not all—of the time.

Cosmic Surprises, Then a Pattern

Could physicists detect those pions and confirm Yukawa's bold prediction? The tools for detecting subatomic particles were getting better, and scientists began looking for pions in cosmic rays, the stream of high-energy particles striking Earth from space. Another possible source of pions was in powerful new machines that could accelerate particles to

New Particles Made Here. This photograph shows the inside of the linear accelerator at Fermilab near Chicago, where subatomic particles undergo high-energy collisions. This has led to the discovery of many previously unknown members of what physicists at one time called the particle zoo.

very high energies. In 1937, cosmic ray particles were discovered that had about the expected mass of a pion but did not respond to the strong force. "Who ordered that?" asked a famous physicist Isidore Rabi about the particle we now call a muon, which seemed to be nothing more than a super-sized electron.

With improved detectors, in 1947 cosmic ray physicists finally found pions and their super-sized cousins, kaons, in cosmic rays, and that was only the beginning of the story.

By the end of the 1950s, a whole "zoo" of subatomic particles with names like lambda, sigma, and xi had been discovered in cosmic rays and particle accelerators. In 1962, physicists even found a second neutrino, which they named the muon neutrino since it appeared to be a partner to the muon in the same way that the renamed electron neutrino was to the electron. As the number of subatomic particles grew, physicists were faced with a question similar to the one chemists struggled with a century earlier as more elements were discovered. They noticed hints of patterns in the properties of these new particles, but they couldn't figure how to classify them. That problem was solved by Murray Gell-Mann (1929–).

In 1954, Gell-Mann and Japanese physicist Kazuhiko Nishijima noticed that the interactions of these new particles seemed to be obeying conservation laws not only for known quantities (such as energy, charge, and spin) but also for a quantity that had no clear physical significance. Gell-Mann called that quantity strangeness. In 1961, Gell-Mann and amateur physicist Yuval Ne'eman, a colonel in the Israeli army, each recognized that these particles had properties that fit a certain mathematical pattern. The symmetries in this pattern suggested the existence of another particle. Gell-Mann described its properties, including its negative charge. Because it would complete the pattern, he named it the omega-minus after the last letter of the Greek alphabet. It was discovered in 1964.

Gell-Mann went on to make a surprising statement about matter: protons, neutrons, and strange particles are not fundamental but are each composed of three smaller entities that he named quarks. The strong interaction was not between protons and neutrons, but between the quarks that composed them. Quarks came in three "flavors," which he designated up, down, and strange, and they interacted with each other by exchanging particles called gluons. Meanwhile several physicists were

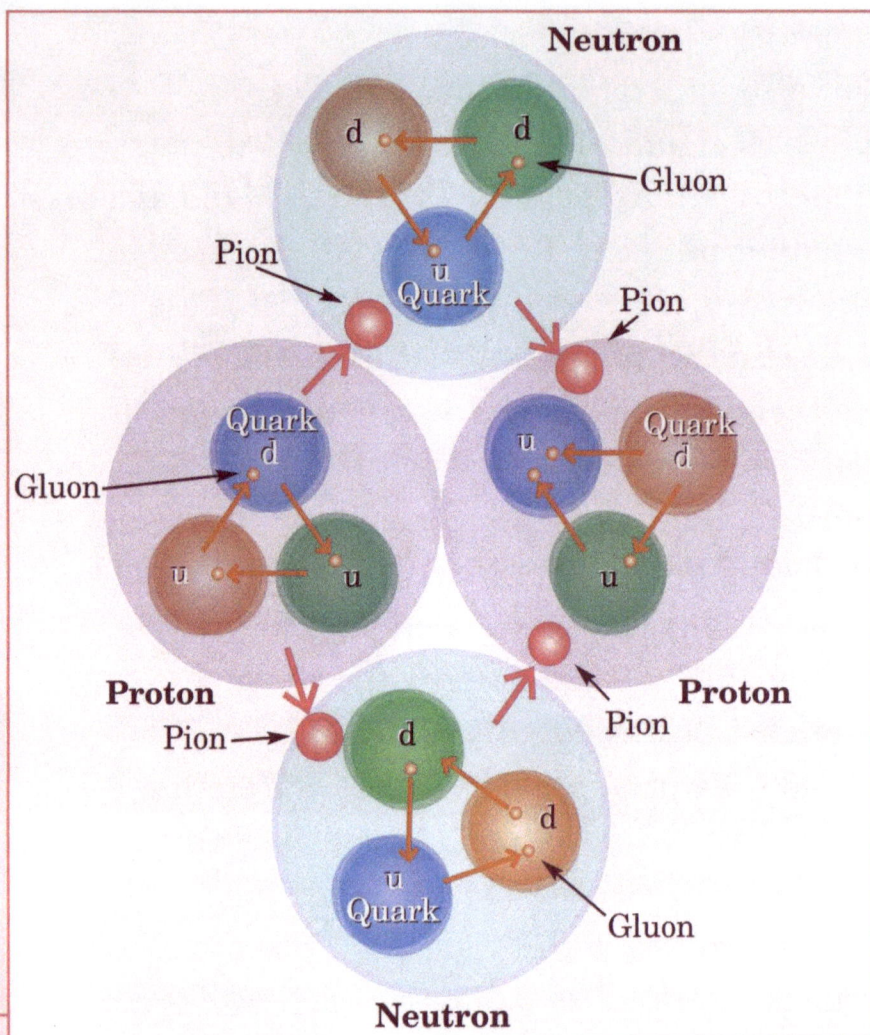

Inside the Alpha Particle. This diagram shows how the helium nucleus, or alpha particle, is held together by the strong nuclear force. The alpha consists of two protons and two neutrons, each made of three quarks, one of each "color." Besides color, quarks also have "flavors," such as "up" and "down." Protons contain two up quarks and one down quark, while neutrons have two downs and one up. Within each proton or neutron, the quarks attract each other by exchanging gluons. The protons and neutrons attract each other by exchanging pions, which are made of a quark and an antiquark of the corresponding anticolor.

working on a new theory that combined electromagnetism and the weak force into what became known as the electroweak force. Their theoretical work suggested that nature might have another quark, called charm.

By 1974, the list of basic subatomic particles consisted of four quarks, the electron, and the muon plus their corresponding neutrinos and the particles that carried the electroweak and strong forces. Then the next year, along came the tau, an even more massive version of the electron and the muon, and with it the prediction of two more quarks and another neutrino. Was physics heading for yet another explosion of new particles? When would it end? The surprising answer came from huge underground neutrino detectors, and it is the subject of the closing chapter of our story.

Chapter Four

How Many Neutrinos Are There?

The early inhabitants of the particle zoo were discovered in cosmic rays, but since the mid-1950s, all newly discovered particles have been created in huge machines. Those machines accelerate protons, electrons, and other subatomic particles—sometimes even large nuclei—to high velocities, where they have enormous kinetic energy, then smash them into other particles to see what comes out as some of their energy is transformed into mass.

Charged particles are much easier to detect than neutral ones. As they pass through matter, they knock loose electrons from atoms and produce a trail of electrically charged ions that can be detected in a number of ways. For example, if they pass through a vapor just on the edge of condensing into a liquid, a streak of cloud condenses on the ions.

That effect is similar to the way high-flying airplanes produce streaks in the sky. Cloud chambers, used as early as the 1890s for studying the ionization due to X-rays and radioactivity, were the earliest devices used to detect particles in cosmic rays around 1910.

Many other particle detection techniques have since been developed. They employ strong magnetic fields, so charged particles passing through them follow circular arcs. Positive charges curve in one direction and negative charges curve in the opposite. The radius of the curve enables the experimenters to determine the particle's momentum. If they can determine its energy (which they often can from characteristics of the trail or other knowledge), that gives them the particle's mass. Knowing mass and charge is usually enough to identify exactly which member of the particle zoo the detector has caught. Or it may indicate the discovery of a previously unknown particle.

Particle physicists (and their computers) have to plow through enormous amounts of data to catch a glimpse of rare subatomic particles and events. That mass of information explains why

Solar Neutrino Detector. Deep underground and thus shielded from cosmic rays, the Japanese Super-Kamiokande detector is lined with light-sensing elements that look for the telltale flashes of light that announce the end of the brief life of positrons created in the interaction of electron neutrinos from the Sun with protons in a huge vat of cleaning fluid. When a positron meets its anti-particle, an electron, and the mass of the pair becomes energy, the interaction creates two photons of a particular frequency that go off in opposite directions.

in 1974, while they were celebrating the discovery of the first particles known to contain charm quarks, they hadn't quite noticed the hints of another surprise in their experiments.

Catching Neutrinos

The reason for the celebration was that the discovery of the charm quark seemed to tie up many loose ends. Matter was apparently built from four quarks, which responded to the strong nuclear force, and four "leptons" (the electron, the muon, and their matching neutrinos) that did not. But the celebration was cut short in 1975 when physicists once again had to ask, "Who ordered that?" A new lepton appeared in their experiments. It behaved in the same way as the tiny electron, but its mass was nearly 4,000 times as great. They called it the tau particle, and they presumed it would have a matching neutrino as well—which seemed to call for two more quarks, named top and bottom (or more colorfully, truth and beauty). Between 1995 and 2000, particle physicists found evidence of the two new quarks and the tau neutrino, but they couldn't help wondering if other generations of particles remained to be discovered. By 2002, nature was providing hints that the answer was no. To read those hints, scientists had to gather neutrinos from sunbeams.

Physicists detect subatomic particles by the way they interact with matter through electromagnetic and nuclear forces. The three types of neutrino carry no electric charge, so they don't interact electromagnetically, and as leptons, they don't interact through the strong force. The weak force, to which neutrinos will respond, keeps particles like neutrons together, but not perfectly. The most likely way to detect a neutrino would be in a rare process of reverse beta decay, in which the weak force between a very energetic neutrino and a proton produces a neutron and a positively charged version of the electron called a positron.

To catch a neutrino in this way would require either a lot of energetic neutrinos or a lot of protons. In 1956, two physicists at Los Alamos Scientific Laboratory in New Mexico set out to do the hardest experiment they could think of: neutrino detection. Nuclear reactors produce lots of neutrinos, so they built a detector that could operate in that intense environment. They detected the neutrinos indirectly from flashes of light produced by the positrons in a detector, confirming that neutrinos really exist. Meanwhile, astrophysicists had been developing theories about the way the Sun and other stars produce their energy by combining light nuclei (like hydrogen) to form heavier ones (like helium), with some mass being transformed into energy, a process called nuclear fusion. The theories were quite successful in explaining the behavior of the stars and the mixture of different atoms in the universe. They also predicted precisely the number of neutrinos streaming from the Sun to Earth every second.

Neutrinos from the Sun are more spread out than neutrinos from a nuclear reactor, so to test the theory, scientists needed lots of protons to catch the neutrinos and a detector that would respond to a reverse beta decay. Cosmic rays might produce similar signals, so the device had to be placed far underground where neutrinos were the only particles that would penetrate. The first such device, a giant vat of cleaning fluid surrounded by instrumentation, went into operation in the Homestake Gold Mine in South Dakota in the 1980s, just in time to detect neutrinos from a giant supernova explosion in a neighboring galaxy in 1987. However, it was detecting too few neutrinos from the Sun. Larger neutrino detectors were built in Japan and Canada, and they, too, found too few solar neutrinos—only about one-third the expected number. Was something wrong with the theory

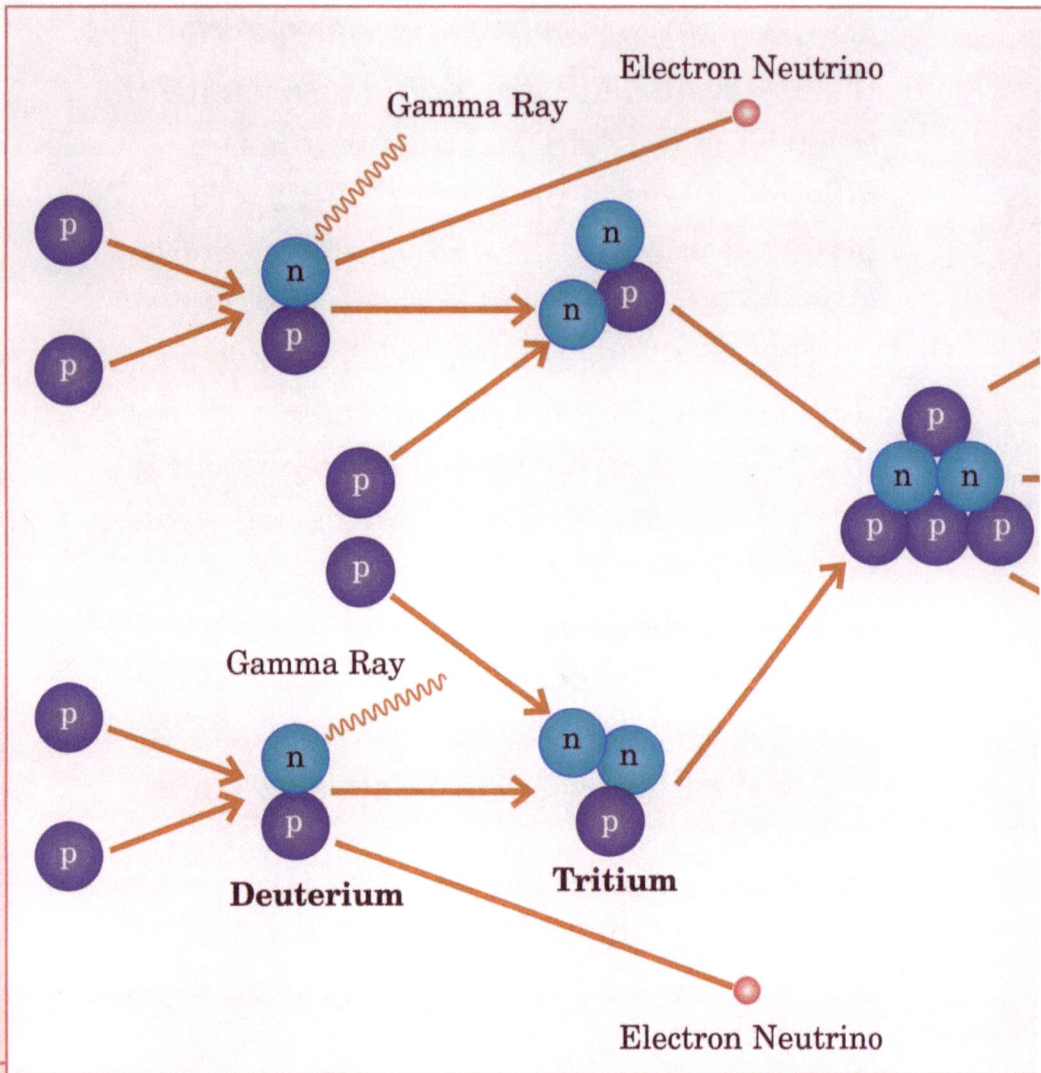

Electron Neutrino

Gamma Ray

p

n

p

p

n

n

p

p

p

n

n

p

p

p

Gamma Ray

p

n

p

Deuterium

n

n

p

Tritium

Electron Neutrino

Inside the Neutrino Factory. This diagram shows the nuclear fusion process by which the Sun makes its energy, creating neutrinos as well as light. Many different fusion sequences take place, but the first step is the most important source of solar neutrinos. In that step, two protons join to produce a deuterium nucleus (a heavier form of hydrogen containing a proton and a neutron). To conserve electrical charge, energy, momentum, and spin, the fusion reaction also produces a positron and a neutrino.

Helium

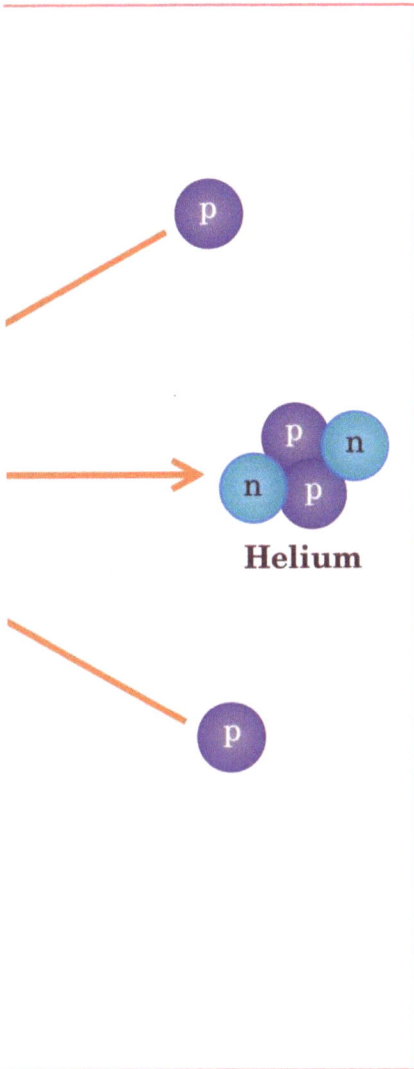

of the Sun's nuclear fusion, was something wrong with the detector, or was something wrong with the theories about neutrinos and their interactions with matter?

Particle theorists had an unusual suggestion. Their equations were telling them that the three different kinds of neutrino might not be different after all. As a neutrino moves through space, it might repeatedly change from an electron neutrino to a muon neutrino to a tau neutrino over and over again, a process referred to as oscillation. After the long trip from the Sun, the neutrinos would arrive as an even mix of all three types. In 2002, experimenters at the

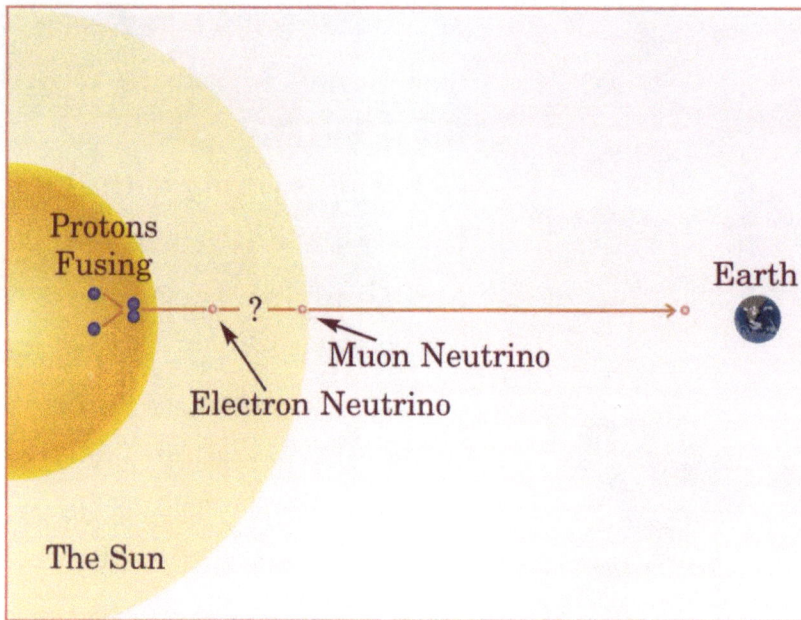

The Mystery of the Disappearing Neutrinos. Physicists who study the Sun were surprised by the low rate of neutrinos captured by Super-Kamiokande and other neutrino detectors. The measurements were approximately one-third of the expected values. Particle theorists came up with a very satisfying explanation. The three forms of neutrinos—electron, muon, and tau—are not different particles but rather different "modes" of the same particle. Solar neutrinos are created in the electron neutrino mode, but by the time they reach Earth, they are mixed equally among the three modes. Only the electron neutrinos interact with the cleaning fluid, so only one-third of the neutrinos are detected.

Sudbury Neutrino Observatory in a nickel mine deep beneath Ontario, Canada, reported evidence that oscillation was indeed taking place.

That means there aren't three kinds of neutrino, but only one type with three different modes. Since the solar neutrino detectors only

Quarks, Leptons, and Their Associated Neutrinos

Up + Down Quarks	Electron	Electron Neutrino
Strange + Charm Quarks	Muon	Muon Neutrino
Bottom + Top Quarks	Tau	Tau Neutrino

pick up neutrinos in their electron mode, they would only register a third of the expected number, which meant that the measurements matched the theory after all. If matter had a fourth set of basic subatomic particles, then the neutrino would have four modes and even fewer would be detected. So the number of solar neutrinos detected points to a limit to the number of new subatomic particles.

Neutrinos may also offer clues to other cosmic mysteries that have emerged in the late twentieth and early twenty-first centuries. For instance, the way galaxies spin indicates that they have much more mass than their visible parts would suggest. Cosmologists, scientists who study the history of the universe, are puzzled about all the cold dark matter that seems to exist in and between galaxies. Might some of that come from neutrinos? Perhaps. The best scientific guess is that the total mass of the protons, neutrons, and electrons in the universe is matched by approximately an equal mass of neutrinos. It's not enough to account for all the undetected mass, but the detection and study of neutrinos is still in its early stages. That means it's likely to produce many more surprises. That little neutral particle first rescued the principles of conservation of momentum and energy in beta decay. Then its ability to change modes told scientists that their present list of basic subatomic particles may be all that the universe has to offer. Who knows what other discoveries the neutrino will lead us to—if only we can figure out how to follow it!

Glossary

alpha particle or alpha ray A helium nucleus that is emitted from some radioactive elements.

atomic mass or atomic weight A number that specifies the mass of the atoms of a particular element. For naturally occurring elements, it is approximately equal to the number of protons plus the average number of neutrons in the nuclei.

atomic number The number of protons in the nucleus of an atom, which determines its chemical identity as an element.

beta particle or beta ray An electron that is emitted from some radioactive elements.

chemistry The science of the properties of matter based on interactions of atoms and molecules.

compound A substance made of only one kind of molecule that consists of more than one kind of atom. For example, water is made of molecules that contain two atoms of hydrogen and one atom of oxygen.

cosmic rays High-energy particles that stream through space and are detected on Earth.

electromagnetism A fundamental force of nature, or property of matter and energy, that includes electricity, magnetism, and electromagnetic waves, such as light.

electron A very light subatomic particle (the first to be discovered) that carries negative charge and is responsible for chemical properties of matter.

element A substance made of only one kind of atom.

gamma ray A high-energy photon that is emitted from some radioactive elements.

gluon A particle that is exchanged between quarks, resulting in their being bound together.

ionization A process in which neutral atoms are turned into electrically charged ions by gaining or losing electrons.

lepton A subatomic particle that does not respond to the strong nuclear force. The leptons include electrons, muons, taus, and their corresponding neutrinos.

molecule The smallest bit of matter that can be identified as a certain chemical compound.

muon A lepton first discovered in cosmic rays that is about 250 times as massive as an electron.

neutrino A subatomic particle with very little mass and no electric charge that is emitted along with an electron in beta radiation.

neutron A subatomic particle with neutral electric charge found in the nucleus of an atom.

nuclear fusion A process in which lighter nuclei combine to form heavier nuclei and release energy. This process powers the stars and produces neutrinos that stream outward into space.

nucleus The very tiny, positively charged central part of an atom that carries most of its mass.

periodic table of the elements An arrangement of the elements in rows and columns by increasing atomic number, first proposed by Dmitry Mendeleyev, in which elements in the same column have similar chemical properties.

photon A particle that carries electromagnetic energy, such as light energy.

physics The science of matter and energy.

pion A subatomic particle that carries the strong nuclear force, binding protons and neutrons together in the nucleus.

proton A subatomic particle with positive electric charge found in the nucleus of an atom.

quantum electrodynamics (QED) A mathematical description of the electromagnetic force that accounts for quantum mechanical phenomena.

quantum mechanics A field of physics developed to describe the relationships between matter and energy that account for the dual wave-particle nature of both.

quantum number One of several numbers that specifies the state of a property of a subatomic particle, such as its orbital characteristics within an atom or its spin.

quark A sub-subatomic particle that exists in several forms that combine to make protons, neutrons, and some other subatomic particles.

radioactivity A property of unstable atoms that causes them to emit alpha, beta, or gamma rays.

scattering An experimental technique used to detect the shape or properties of an unseen object by observing how other objects deflect from it.

spin A property of subatomic particles expressed by a quantum number that describes the way an electron may align in a magnetic field.

strong nuclear force or strong force A fundamental force of nature that acts to hold the protons and neutrons in a nucleus together.

tau The largest lepton, having about 4,000 times the mass of an electron.

theory of relativity A theory developed by Albert Einstein that deals with the relationship between space and time. Its most famous equation ($E = mc^2$) described the relationship between mass and energy.

transmutation A transformation of the nucleus of one element into another by emission of an alpha or beta particle.

weak nuclear force or weak force A fundamental force of nature that is responsible for beta decay of a radioactive nucleus.

• For More Information

Organizations

Lederman Science Center
Fermilab MS 777
Box 500
Batavia, IL 60510
Web site: http://www-ed.fnal.gov/ed_lsc.html
This museum is an outstanding place to discover
the science and history of subatomic particles. It
is located at the Fermi National Accelerator
Laboratory (Fermilab) outside of Chicago.

Magazines

American Scientist
P.O. Box 13975
Research Triangle Park, NC 27709-3975
Web site: http://www.americanscientist.org

New Scientist (U.S. offices of British magazine)
275 Washington Street, Suite 290
Newton, MA 02458
Web site: http://www.newscientist.com

Science News
1719 N Street NW
Washington, DC 20036
Web site: http://www.sciencenews.org

Scientific American
415 Madison Avenue
New York, NY 10017
Web site: http://www.sciam.com

Web Sites

Due to the changing nature of Internet links, the Rosen Publishing Group, Inc., has developed an online list of Web sites related to the subject of this book. This site is updated regularly. Please use this link to access the list:

http://www.rosenlinks.com/lsap/neutrino

For Further Reading

Close, Frank, Michael Marten, and Christine Sutton. *The Particle Odyssey: A Journey to the Heart of Matter*. New York: Oxford University Press, 2002.

Cooper, Christopher. *Matter* (Eyewitness Books). New York: Dorling Kindersley, Inc., 2000.

Henderson, Harry, and Lisa Yount. *The Scientific Revolution*. San Diego: Lucent Books, 1996.

Narins, Brigham, ed. *Notable Scientists from 1900 to the Present.* Farmington Hills, MI: The Gale Group, 2001.

Strathern, Paul. *Mendeleyev's Dream: The Quest for the Elements.* New York: Berkeley, 2002.

Tweed, Matt. *Essential Elements: Atoms, Quarks, and the Periodic Table*. New York: Walker and Company, 2003.

Bibliography

Cropper, William H. *Great Physicists: The Life and Times of Leading Physicists from Galileo to Hawking.* New York: Oxford University Press, 2001.

Gell-Mann, Murray. *The Quark and the Jaguar: Adventures in the Simple and the Complex.* New York: W. H. Freeman, 1994.

Kragh, Helge. *Quantum Generations: A History of Physics in the Twentieth Century.* Princeton, NJ: Princeton University Press, 1999.

Nobel Foundation. *Nobel Lectures in Physics 1901–1921.* River Edge, NJ: World Scientific Publishing Company, 1998.

Seife, Charles. *Alpha & Omega: The Search for the Beginning and the End of the Universe.* New York: Viking, 2003.

Young, Hugh D., and Roger A. Freedman. *University Physics: Extended Version with Modern Physics.* Reading, MA: Addison-Wesley Publishing Co., 2000.

• Index

About the Author

Award-winning children's author Fred Bortz spent the first twenty-five years of his working career as a physicist, gaining experience in fields as varied as nuclear reactor design, automobile engine control systems, and science education. He earned his Ph.D. at Carnegie-Mellon University, where he also worked in several research groups from 1979 through 1994. He has been a full-time writer since 1996.

Photo Credits

Cover, pp. 1, 3, 8, 10–11, 24–25, 31, 34, 40, 48–49, 50 by Thomas Forget; p. 7 © Science, Industry, and Business Library/New York Public Library/Science Photo Library; pp. 9, 18 © Science Photo Library; p. 22 © Prof. Peter Fowler/Science Photo Library; p. 29 U.S. Library of Congress/Science Photo Library; p. 37 © Fermilab/Science Photo Library; p. 44 © Los Alamos National Laboratory/Science Photo Library.

Designer: Thomas Forget; Editor: Jake Goldberg

www.ingramcontent.com/pod-product-compliance
Lightning Source LLC
Chambersburg PA
CBHW050910210326
41597CB00002B/86